FIGHTING FORCES ON LAND

M1097 HUMVEE

DAVID BAKER

Rourke
Publishing LLC
Vero Beach, Florida 32964

www.rourkepublishing.com

PHOTO CREDITS: All photos courtesy United States Department of Defense, United States Department of the Army, AM General

Title page: *Military police patrol Iraq in their Humvee.*

Editor: Robert Stengard-Olliges

Library of Congress Cataloging-in-Publication Data

Baker, David, 1935-
   M1097 Humvee / David Baker.
       p. cm. -- (Fighting forces on land)
   Includes index.
   ISBN 1-60044-244-7
   1. Hummer trucks--Juvenile literature. 2. United States--Armed
Forces--Transportation--Juvenile literature. I. Title. II. Series.
   UG618.B355 2007
   623.7'4722--dc22
                                    2006010672

Printed in the USA

CG/CG

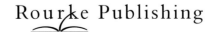

www.rourkepublishing.com – sales@rourkepublishing.com
Post Office Box 3328, Vero Beach, FL 32964

Table of Contents

CHAPTER ONE

THE NEED

To the US Army it is known as the Humvee but the vehicle that has become as much sought after for **civilian** use as for the military is officially named the Hummer. Built by AM General, it has revolutionized the role of the general-purpose vehicle and become as famous as the Jeep all over the world, where it is sold in large numbers.

▲

With light deflector shields protecting the gunner manning a M2 machine gun Marines from the 1st Battalion, 503rd Infantry Regiment patrol the area around Ar-Ramadi, Iraq.

Unlike anything used by the Army before, the Humvee has proved an adaptable all-purpose light vehicle adopted by many armed forces for a broad range of missions and jobs.

▲

The specification for the Humvee was demanding, prompting AM General to design a vehicle from scratch for tough and challenging conditions.

▲

In the busy and congested streets of downtown Baghdad, a Humvee patrol seeks a fine line between trust and suspicion ever conscious of the threat that can challenge everyday activity and trading in the market area.

Its military name is an **acronym**, an abbreviation of High Mobility Multi-Purpose Wheeled Vehicle, or HMMWV. It emerged from a requirement first issued by the Army in 1979 for a vehicle capable of carrying loads of 500 lb to 2,500 lb across difficult **terrain** in punishing weather and harsh climates. Aimed at replacing the M561 Gama Goat and the M-151 series of Jeeps and civilian trucks it was to be an all-purpose vehicle for a variety of tasks and jobs.

CHAPTER TWO

A FRESH START

The competition to build a replacement for a wide range of small, aging utility trucks attracted design concepts from Teledyne, building on its existing Cheetah vehicle, and Chrysler (later sold to General Dynamics), upgrading its Saluki desert design. AM General decided to propose a completely new design from scratch and not to build on anything that had gone before.

▲

This virtual Humvee simulator trains drivers about road conditions and operational tactics.

▲

The distinctive radiator grill and air louver panel is often the only distinguishable feature of this all-purpose vehicle, the rest of its shape transformed by adaptations and modifications and make it the most used military vehicle in the world.

▲

The basic Humvee has a sound design with a rugged and strong chassis capable of carrying several different conversion kits allowing the vehicle to carry weapons, equipment, or people.

This was a smart move because the Army wanted a single vehicle to replace the wide variety of trucks and pick-ups it had used before. The fresh start paid off. When the three contenders went for trials in 1980, the AM General design was a clear winner. Army requirements were extremely demanding, **stipulating** deep water **fording** capability, reliability in freezing Arctic or hot desert conditions, and easy maintenance. Of the three, AM General's design came out as the lightest, strongest, most **adaptable**, and the most reliable. In March 1983 AM General received a five year production order for 55,000 Humvees.

▲

Designed for all weather and any climate on earth, these Humvees push through snow packed passes as they travel roads unsuitable for any other tracked vehicle.

UNDER THE HOOD

Regarded by some in the Army as a "Jeep on steroids," the Humvee is as tough and as rugged as it gets. Powered by a 6.5 liter, V8, **diesel** engine pushing out 160 hp at a sedate 3,400 rpm, the transmission puts power through the four-wheel drive via four forward gears and one reverse. The Army required diesel instead of gasoline because it standardized on this fuel throughout its vehicle inventory in the early 1980s.

▲

Using the hood as a table for their map, airborne troops work with the 155th Brigade Combat team to plan a convoy route for Humvees on their way to inspect construction of the Mishkub police station near Najaf, Iraq.

▲

Massive coil springs visible behind the front wheel display rugged engineering with a strong chassis capable of rough field operation. It exists to carry people and support their mission, in this case a light foot patrol.

▲

A patrol from the 59th Military Police Company in Mosul, Iraq, with a well equipped Humvee watch for suspicious activity. The rear of their vehicle is covered with racks and baggage with ample attachments, points, and covers.

At the curb the Humvee weighs 5,200 lbs empty but it can carry a payload of up to 2,500 lb, although later versions can carry 4,400 lb for a total driving weight of more than five tons. With a low profile – the Humvee is only six feet tall – and with a length of only 15 feet it has a wide stance of seven feet that makes it very stable over rough and inclined terrain.

▲

A convoy of mixed army vehicles, including several Humvees, moving through a village in Kosovo during civil war in the former Yugoslavia.

This is made possible due to the raised **differential** freeing the underside from obstructions that could snag rocks or ground the vehicle. Seating, 1+1 or 2+2 according to the model, straddles the raised drive shaft tunnel giving the vehicle a 16 inch ground clearance. All done, the Humvee is capable of going almost anywhere and is very exciting to drive!

▲

High up in the Mushla mountains in Iraq, a Humvee makes light work of icy roads as it moves forward to patrol remote regions suspected of harboring insurgents and suicide bombers, all in a day's work for vehicles of this type.

On reconnaissance with the Humvee, its 16-inch ground clearance, greater than any other vehicle in its class, gives it good traction on rough surfaces.

CHAPTER FOUR

★ DOWN AND DIRTY

In the Army, the basic Humvee, the M1097, comes with an open rear section and a roll bar with fixed frame windshield. When used as a troop carrier it has an optional hardtop and seating for ten people including the driver. As a shelter carrier it doubles as a hard shell truck with a carrying capacity in excess of two tons.

▲

Humvees frequently set up roadblocks or security points carrying a squad to any location at speed.

▲

Responding rapidly to a suicide bomber, soldiers from the 3rd Infantry Division set up a security perimeter near the International Zone in Baghdad, Iraq, using their Humvee.

As a TOW (Tube-launched, Optically-tracked, Wire command-link guided) missile carrier, **designated** M1025A2, it has armor protection for getting to the forward areas of battle restocking Bradley and Stryker vehicles with anti-tank weapons. The Army did not require the Humvee to have basic armor protection but a special package of steel, **Kevlar,** and layered polycarbonate windows can upgrade the vehicle to survive small arms fire, shrapnel, and low-yield fragmentation mines.

▲

The 0.5 machine gun is a powerful weapon and the highest caliber carried as standard on Humvee variants and derivatives.

With a 7.62 mm machine gun for defensive purposes, troops search for hidden ordnance somewhere in Iraq.

▲

The Humvee makes an ideal ambulance and in various conversions, it serves as a medical evacuation truck able to collect wounded and speed them to the hospital.

There are 18 basic variants of the Humvee in service with US armed forces, optional equipment can transform the weapons carrier version into a grenade launcher, and it can carry a mounted M2 machine gun for protection. Other models serve as ambulances and communications vehicles, even supporting artillery as a fire control vehicle.

▲

Equipped with a M2 machine gun in a hardtop mount the Humvee provides light backup to infantry patrols serving in a peacekeeping role.

CHAPTER FIVE

ADDED PROTECTION

Improvements to the basic Humvee have carried it into many different roles. In 1995, it was selected as the **chassis** for a scout, military police, and bomb disposal vehicle. Protected against 7.6 mm bullets, 155 mm artillery airbursts, and 12 lb anti-tank mines, the M1114 model carries the Humvee into places the basic vehicle fails to reach.

▲

Humvees on convoy protection duty near a power station in Iraq. With winch and hawser the vehicle can be used to pull other disabled light vehicles or haul road debris away to clear access roads.

▲

Humvee was designed from the outset to accommodate a wide range of weapons including small arms and light to medium machine guns, important capabilities for such an all-purpose vehicle.

▲

US and Albanian troops form an international police keeping crew aboard this Humvee in Kosovo giving a clean shot of its low profile and wide seat spacing just discernible through the windshield.

With a light hard-shell cover, this Humvee straddles river rocks as it negotiates a stream. ▶

Beyond these improvements, other countries have bought the Humvee and adapted it to their own unique requirements. The Swiss firm Mowag has produced the lightweight armor-plated Eagle that gives its occupants protection from nuclear, **biological**, and chemical contamination with an optional assortment of armaments and manually, or remotely, operated turrets.

Turkey has gone even further and built up the basic chassis. Their version has full armor protection, room for up to 11 people, and carries a wide range of weapons. Known as the Cobra it even has a swim kit transforming it into a fully amphibious vehicle.

CHAPTER SIX

STREETWISE

The Humvee is the most **prolific** military vehicle in the world with more than 10,000 used by US forces during Operation Iraqi Freedom in 2003. More than 190,000 have been produced by AM General and other manufacturers under license.

▲

The low profile of the Humvee can be used to advantage when cover is needed to remain hidden.

▲

Collapsed side plates flat against the roof expose the M2 machine gun and ammunition box on this patrolling Humvee in Iraq.

No mention of the Humvee can avoid its wide appeal to people everywhere, making it one of the nation's most desired off-road vehicles. Customized Hummers add color to plush interiors and monster truck versions cruise the streets, go off road or race each other on track.

With a name as famous as the universal Jeep, Hummer has entered the language of more countries around the world than any other light truck. The M1097 Humvee serves many militaries around the globe and will be tough act to follow.

Glossary

acronym (AK ruh nim) – a word made by the first or first few letters of the words in a phrase

adaptable (uh DAPT uh buhl) – to be able to change to suit

biological (bye ol LOJ ikal) – organisms or toxins found in nature

civilian (si VIL yuhn) – someone who is not a member of the armed forces

chassis (CHASS ee) – the frame on which the body of a vehicle is built

designated (DEZ ig nate ted) – named or called something

diesel (DEE zuhl) – a fuel used in diesel engines that is heavier than gasoline

differential (DIF ur uhnt shuhl) – gears that supply equal force to the driving wheels

fording (for DING) – crossing a shallow stream or river

kevlar (KEV lar) – a synthetic material which is lightweight, flexible , and five times stronger than steel. It is often used as armor for bullet-proof vests

prolific (pruh LIF ik) – very productive or producing a large quantity

stipulating (STI puhl ating) – to specify as a requirement of an agreement or contract

terrain (tuh RAYN) – the physical features of the land or ground

INDEX

FURTHER READING

Monro, Bill. *Humvee*. Crowood Press, 2002
Healy, Nick. *High Mobility Vehicles: The Humvees*.
     Capstone Press, 2005
Lamm, John and Delorenzo, Matt. *Hummer H2*.
     Motorbooks International 2003

WEBSITES TO VISIT

http://en.wikipedia.org/wiki/Humvee
http://www.amgeneral.com/vehicles_hmmwv.php

ABOUT THE AUTHOR

David Baker is a specialist in defense and space programs, author of more than 60 books and consultant to many government and industry organizations. David is also a lecturer and policy analyst and regularly visits many countries around the world in the pursuit of his work.